How To Raise Strong And Healthy Farm Animals

3 books in 1

Copyright © 2016 HTeBooks

Copyright © 2016 HTeBooks

All rights reserved. This book or any portion thereof may not be reproduced or used in any manner whatsoever without the express written permission of the publisher except for the use of brief quotations in a book review.

Disclaimer

This book is designed to provide condensed information. It is not intended to reprint all the information that is otherwise available, but instead to complement, amplify and supplement other texts. You are urged to read all the available material, learn as much as possible and tailor the information to your individual needs.

Every effort has been made to make this book as complete and as accurate as possible. However, there may be mistakes, both typographical and in content. Therefore, this text should be used only as a general guide and not as the ultimate source of information. The purpose of this book is to educate.

The author or the publisher shall have neither liability nor responsibility to any person or entity regarding any loss or damage caused, or alleged to have been caused, directly or indirectly, by the information contained in this book.

Table of Contents

HOW TO RAISE STRONG & HEALTHY DUCKS 6

WHY RAISE DUCKS? 7

STARTING OFF 12

HATCHING THE EGGS 16

TAKING CARE OF THE DUCKLINGS 19

RAISING HEALTHY DUCKS 25

HOW TO APPLY WHAT YOU'VE LEARNED? 31

HOW TO RAISE STRONG & HEALTHY PIGS 32

STARTING OUT 33

HOUSING YOUR PIGS 40

FEEDING YOUR PIGS 45

HYGIENE 48

HOW TO RECOGNIZE DISEASE IN YOUR PIG FARM 51

HOW TO APPLY WHAT YOU'VE LEARNED? 55

HOW TO RAISE STRONG & HEALTHY CHICKENS .56

BEGINNING AT THE BEGINNING: CHOOSING THE TYPE OF CHICKEN TO RAISE 57

PREPARING A HOME FOR YOUR CHICKS 60

HOUSING: THE CHICKEN COOP & CHICKEN RUN 66

FEEDING IDEALS FOR STRONG AND HEALTHY LAYERS: IDEALS ON WHAT, HOW AND WHEN TO FEED YOUR CHICKEN AT DIFFERENT AGES AND STAGES AS WELL AS WHAT THEIR FEED ENTAILS.. 70

FEEDING IDEALS FOR STRONG AND HEALTHY BROILERS: IDEALS ON WHAT, HOW AND WHEN TO FEED YOUR CHICKEN AT DIFFERENT AGES AND STAGES AS WELL AS WHAT THEIR FEED ENTAILS 76

SAFETY & HEALTH: FOCUS ON THE MOST COMMON YET LETHAL OF ILLNESSES, THEIR SYMPTOMS, PREVENTION AND THEIR TREATMENT 79

DIFFERENT CHICKEN BREEDS: WHAT TO KNOW ABOUT VARIED BREEDS BEFORE SELECTION SO AS TO BOOST YOUR CHANCES OF OWNING STRONG AND HEALTHY CHICKEN 86

HOW TO APPLY WHAT YOU'VE LEARNED? 93

How To Raise Strong & Healthy Ducks

Have you been toying with the idea of rearing birds to have fresh supply of poultry meat but have been unable to do that because you don't have the right information on how to go about it? Have you considered rearing ducks? Well, if you've not yet settled on any particular bird, then you should probably try ducks and you'll never regret it.

Whether you are a beginner or you already have some prior experience in keeping ducks, you will have to know how to raise strong and healthy ducks. So many people often give up on ducks before they can even succeed especially after discovering that rearing ducks is different from chicken.

If you want to start a journey that you won't quit, this book will help you to do just that. It will guide you through the entire process in details so that you know what to do at what time when raising the ducks. If you follow the steps as they are explained, then there would be no reason for failure in your journey.

Why Raise Ducks?

"Take advantage of it now, while you are young, and suffer all you can, because these things don't last your whole life."

- Gabriel García Márquez,

If you are still not sure of what you really want to rear, perhaps you are torn between rearing ducks and any other bird. As such, it is probably best that we start by discussing why you should raise ducks in the first place.

The nutritious eggs

A duck's eggs have very high nutritional content compared to that of a chicken. The eggs are very high in vitamin A, D and Omega-3 fatty acids. They are also alkaline in nature hence when consumed; they create an acidic environment in your body, which has plenty of health benefits. When you bake duck eggs, they come out way richer and fluffier. They are also bigger compared to a chicken's eggs. The eggs are delicious too especially when you soft-boil them. Generally, you can take them as you would a normal chicken egg.

The eggs are also perfect for baking since they have lower water content and higher fat.

The shells are thick and are used for art especially by the eastern European artists that make fine pysanky. This means that you can make money out of them by selling them or make some art using them. You can also sell the eggs at a higher price than chicken eggs.

The eggs also have longer shelf life because of their hard shells and membranes.

They do not require a lot of maintenance

You can feed the kind of food fed to chicken to your ducks. The best thing about them is that the higher percentage of their food comes from their foraging and grazing. This means that you will have less work feeding them. Ducks do not require a big space to stay too.

They are good for pest control

Ducks are good hunters of annoying pests such as slugs, worms, spiders, grasshoppers, flies, crickets, grubs and snails, which may destroy your plants. They might even eat mice, toads and small snakes. You can let them into your shade garden to protect your garden from pests and as bonus, feed. Unlike chicken, ducks don't scratch the ground for their food hence they won't disturb your plants and upend your garden. Just be careful as they like eating ripe strawberries and lettuces. You should also not let them in a garden with seedlings, which are rather fragile as they might destroy them by trampling them since ducks have large flat feet.

Ducks are more resistant to cold

These birds have an extra layer of fat which chicken do not have. They also have waterproofed feathers, which keeps them dry and warm when in water. This therefore makes ducks more resistant to cold than chicken.

Ducks have a higher tolerance to heat

When it is hot, you may notice that chicken usually stand around as they pant or hide in the shades. Ducks on the other hand paddle in the pool quite comfortably. The ducks handle the heat easily by just taking a dip to cool off.

Ducks are not noisy

As opposed to chickens, ducks are a lot quieter. They may quack only when they feel excited or agitated.

They lay eggs regularly

When compared to chicken, ducks lay their eggs more regularly even in the winter season when it is usually cold. Most of the duck breeds also have a very low likelihood of going broody hence messing up your egg production like chicken (broodies are the ones that do not lay eggs).

They are generally healthy

Since ducks spend a lot of time in water, they tend to be less susceptible to being infested by mites and other parasites that affect chicken. This is because all parasites that attempt to latch on to a duck would obviously be drowned. Ducks also have stronger immune systems hence they can stay in better health and have a lesser chance of contracting diseases as it is the case with chicken.

Ducks as pets

Ducks, especially ducklings are very adorable to have as pets as opposed to chicken, which tend to be a bit nervous, skittish, and flighty. On the other hand, ducks are calm, funny at times, and very alert. They are also enjoyable to watch.

They are less aggressive to 'new comers'

Ducks welcome newcomers (whether chicken or ducks) faster than chicken do. As you have noticed, chicken usually test the abilities of the new addition to their flock by pecking rigidly, this can cause serious consequences in the end.

Ducks are friendlier to your lawn

As opposed to chicken, ducks do not scratch the lawn to bare dirt. This means that you can have your ducks on your lawn and it will still look good. Chicken usually eat anything green that tries to grow and worse off dig deep depressions in the ground as they take dust baths or as they cool off when it is hot. Chicken may render your lawn barren due to trampling while ducks on the other hand do very little damage.

Meat

You can also keep ducks for their meat, whether for your own consumption or as a way of earning a living. The larger ones meant for meat will not produce as many eggs as the smaller ones. Different breeds vary in tenderness of meat hence it would be advisable to do your research on which breed suits your preference before you start. It takes approximately 2-4 months to raise a duck for their meat, which of course depends on the particular species.

*Key point/action step

Ducks have very many benefits to you as a farmer. If this is not going to make you want to raise ducks, I am not sure what else will.

Starting Off

"The secret to getting ahead is getting started."

-Mark Twain

Plan the number of ducks to start with

When starting, you need to make a few choices. You don't just get any duck egg and start hatching it. You should start by planning on how many ducks you want to keep depending on the space available to do so. For beginners, 2 ducks are perfect to start with. Do not take just one. This is because these birds are quite sociable creatures hence starting with only one is not advisable. It will become lonely; hence, you might find that it follows you around the compound all day for company. Do not start with too many also as they might become unmanageable. Your space might also be too small for them to fit in hence leading to congestion.

Consider the breed

There are quite a number of domestic duck breeds in the US. Depending on your preference and the purpose of keeping the duck, you might have to do research on the type of breed that suits you.

Pets: these ducks do well in smaller spaces, they are docile and friendly. The best breed in this category is khaki Campbell, which can also provide eggs.

Khaki Campbell

Eggs: These ducks are generally leaner and smaller. This breed should be able to convert the energy from food to production of the eggs instead of channeling it to weight gain. Many of the breeds in this category have a tendency of brooding (stop laying eggs to incubate them). The best ducks for this category are the runner duck and the khaki Campbell.

Runner duck

Meat: If you want to raise ducks for their delicious meat then you should consider going for a breed that converts the energy from the food it consumes to fast weight gain. The ducks in this category have a body frame, which supports a larger size. Some of the best breeds that you can keep for their meat include the pekins and the Rouens. In most commercial dark meat farms, the most common breed is the Pekin duck.

Rouens

Pekins

Find a perfect breeder

At this point, you can either choose to start with eggs or young chicks from a breeder. You should do research on some of the best breeders around your area depending on the type of breed you want. If you wish to start with chicks, then you should consider taking two for starters and see how well you can be able to work on them. If you own an incubator, you can get yourself some eggs from the breed you need then hatch them. You can hatch as many as you wish but ensure you only remain with only two for starters. You can sell the rest or give them out.

Take your time as you make these decisions, as they will alter your result. Ensure you choose a breed that you can deal with and learn more about to avoid any inconveniences when raising it.

*Key point/action step

Your decision on which breed to settle for will largely depend on why you want to keep ducks. However, you should note that ducks are social birds so you need to get at least a pair if you want them to have a good start. Nevertheless, don't start with too many otherwise they might be too hard to manage.

Hatching The Eggs

"Faith is putting all your eggs in God's basket, then counting your blessings before they hatch"

Ramona C Carrol

After choosing the type of breed you want, you should prepare yourself mentally to ensure you are dedicated to the whole process from hatching the eggs, until you have a strong and healthy, full-grown duck.

1. Plan ahead of time

You should keep in mind that a duck's eggs take approximately 28 days to hatch though it might extend to 35 days for other breeds. You need to ensure that the incubator you will be using is prepared prior to purchasing the eggs. You should also remember that the eggs are generally larger than those of chicken; hence, most of the chicken egg incubators will be unable to accommodate duck eggs. Ensure that the tray you use in the incubator is large enough for the duck eggs.

2. Give the incubator time to stabilize

Before you place your eggs inside the incubator, give it some time first so that it can stabilize temperatures. Use a temperature of about 37.5 degrees Celsius and a relative humidity of 55 percent or 29 degrees Celsius if you are using a wet bulb thermometer. As for the ventilation, you should set it as per the manufacturer's instructions. Give a space of up to a full day or two before you can add in the eggs.

3. Select the eggs you want to use carefully

The eggs that you choose must be in good shape so that they can hatch. Avoid using eggs that are cracked, misshapen, double yoked, undersized, oversized, or even dirty. For best results, the eggs should be put in the incubator one to three days after they were laid.

4. Check the incubator regularly

You should check the incubator at least four times each day after placing the eggs inside. Each time you check them, you should turn them over to allow fair distribution of heat on all sides. On the first day of incubation, you should check the eggs after every hour.

5. After one week, check for unfit eggs

After one week has elapsed after incubation of the eggs, you should check for any signs of unfit eggs. Watch out for signs such as clear shells (means that the egg is infertile hence cannot hatch) and eggs that have a clouded shell (they are dead inside). Remove these two types of eggs and leave the rest in the incubator to continue.

6. Transfer the eggs to hatching trays

After about 25 days, you should transfer the eggs to hatching trays. You can either decide to change the setting of your incubator so that it can accommodate the eggs as they hatch or move them to a separate hatching machine. Set the temperature at 37.2 degrees Celsius and a humidity of 65 percent. When you notice the eggs start to pip or any other noticeable change, increase the humidity to reach 80% then increase the opening for ventilation by 50%.

During the last 6 – 12 hours just about when the hatch ends, decrease the temperature to about 36.1 degrees Celsius and change the humidity to 70%. Open the ventilation all the way.

7. Remove the ducklings

Once the ducklings have hatched, remove them from the Hatcher. After about 90 – 95 percent of the ducklings have dried and hatched, you should proceed on to remove them and transfer them to a brooder.

*Key point/action step

The process of hatching the eggs is very delicate; hence, you need to be very careful as you do it to get the best out of them. Be keen especially with the temperatures to avoid damaging the eggs.

Taking Care Of The Ducklings

"It is the time you have wasted for your rose that makes your rose so important."

-Antoine de Saint-Exupéry"

If you opted to go with buying ready ducklings or you hatched your own, this chapter will show you how to take care of the ducklings so that they are healthy and strong.

When the ducklings have hatched from their shells, they require a safe and warm environment so that they can grow well. You need to give them an environment that is free from hazards and give them plenty of water and food. If you are able to provide all these, your playful ducklings will swim and waddle on their own even before you know it.

For this reason, you should know how to feed a duckling well, make them feel at home and keep them away from anything that may harm them. Ensure you remain with 2-4 ducklings since caring for a smaller number is easier especially if you are new to duck keeping. As stated earlier, you should make sure that you have more than one to prevent loneliness since ducks need to socialize with others.

In the case where you bought the ducklings, you might find that most hatcheries sell a minimum of 10-15 ducklings. This number can be overwhelming for you. You should consider giving/selling the extras to responsible relatives and friends.

Make the ducklings feel at home

After the hatching, give them 24 hours so that they can get used to their new environment. During this period, find a brooding box to keep the baby ducks in. You can use a spare bathtub, plastic tote, cardboard box, dog crate, a plastic storage container or maybe a large glass aquarium, which will all serve the same purpose. Don't use a box that has too many holes on the bottom or sides.

Insulate the box well since the ducklings need warmth owing to the fact that their feathers have not yet developed. Use old towels and wood shavings to line the bottom of the box. Avoid using any slippery material such as newspapers since the ducklings still have wobbly legs during the first few weeks of hatching. This means that they might slip on slippery surfaces and hurt themselves.

Use a brooding lamp

The ducklings need to be kept warm enough during the first few weeks after hatching. This is done to allow them to adapt to the chilly air outside the warmth of their eggs. You can use a brooding lamp to achieve this. Buy one from a hardware store or a feed store and install it at the top of your brooder box.

You can start with a 100 watt bulb, which should be enough to provide the required warmth for the young ducklings. Ensure that part of the brooder is placed away from the heat so that the baby ducks can have a place that they can cool off when they need to. Ensure that the bulb is not too close to the baby ducks. This is because they might overheat if they happen to touch the bulb. If the brooder you are using is shallow, prop your lamp higher by using a sturdy prop such as blocks of wood.

Change the placement of the brooding lamp

Check the position of the lamp regularly and adjust it accordingly to ensure that your ducklings are getting just the right amount warmth. You should alter the wattage and heat of the brooding lamp in accordance to the behavior of the ducklings, as they grow older. If you notice that the ducklings are huddling close together under the lamp, you should note that they are cold hence you may need to move the lamp closer, or buy a bulb of higher wattage. On the other hand, if they are scattered out in the brooding box and breathing heavily, then they are probably getting overheated hence the need to move the lamp farther away or use a bulb with a lower wattage.

As the ducks grow older, they need less heat since their feathers will have developed. Raise the lamp or use a lower wattage bulb when you notice that they are not lying under it.

Providing food and water

Provide your ducklings with a lot of water (room temperature). You can do this by putting a shallow bowl or a chick fountain inside the brooding box/brooder, which is just deep enough for the ducklings to dip their beaks rather than their entire heads. The water should be 6.35mm (¼ of an inch) deep at most. This is because ducklings like to clear nostrils as they drink. Giving them a deep bowl full of water would be dangerous since they might climb inside and drown. If you think the drinking bowl is deep for your tiny ducklings to drink from in a safe manner, you should consider lining the bottom of the bow, with marbles or pebbles, which will make it safer for them to drink from.

Water is very important as it generally makes it easier for the ducklings to swallow their food and clean their beak vents. A one-week-old duckling usually consumes about 2L of water each week. At the age of 7 weeks, the duckling will drink about 2L each day.

Feed starter crumbs to your ducklings. For the first 24 hours after hatching, the ducklings will not be eating since they are still absorbing some of the remaining nutrients from the yolk inside the eggs that they hatched from. After that period, you can now start feeding them starter crumbs, which are very tiny pellets of duck food that you can acquire from any local feed supply stores. Buy a plastic feeder and fill it up with the crumbs then place it in the brooder.

If the ducklings hesitate to eat, add a little water to the food to make it easier for them to swallow it. You can also give them sugar water for the first few days to hydrate them and give them energy. To make the right sugar water for your ducklings, just use 80ml (third of a cup) of sugar for every 4L (1 gallon) of clean water. If some ducklings don't drink the water, dip the tips of their beaks into sugar water or shallow water to hydrate them.

If you notice signs of weakness on some ducklings, feed them with duck egg yolk. Normally, weak ducklings may need to be fed a little extra yolk for its nutrition before they can be able to eat starter crumbs. Feed them with some mashed yolk from a duck egg until they start eating starter crumbs.

You should ensure that the ducklings have constant access to food. This will enable them to eat at any time when they feel hungry since at this stage in life they grow quickly. They also need some water to help them swallow the food hence the need to keep the water bowl clean and full at all times.

After a period of about ten days, the ducklings are now ready for grower's pellets. They are just the same as starter pellets but bigger in size.

Change to adult food

After about 16 weeks, the ducklings are adults and can be now fed with adult duck food. After some time, you can add in raw oats to their feeds to give them some additional protein. At most, you should balance the ratio of oats and duck feed to 1:3 respectively. You can also add in commercial chick grit so that the ducklings can digest the food better. Make sure that you change the food on a daily basis since it tends to get wet and thus may develop bacteria and mould if allowed to stay. This might make the ducklings sick.

Do not feed adult ducks/ducklings with food that is not made for ducks

Many of the foods that we normally consume such as bread do not provide ducks with the nutrition they need and some of the foods can even make them sick. Even if the ducks show interest in some of the foods, it would be unwise to feed it to them. If you need to feed the duck with snacks, just cut vegetables and fruits thinly. You can also feed them other healthy treats such as grass, dandelion greens, kale, worms, untreated weeds, peas and some moistened oatmeal after some days. Just make sure that their primary meal is duck food.

Don't feed ducklings with food that is meant for chicken. This is because it does not have the right nutrient content for ducklings.

You should also ensure that you do not use medicated feed for your ducklings as it can lead to organ damage.

*Key point/action step

Little ducklings need all your care and attention so that they can survive. Care for them as you would your own kids. Keep in mind that taking care of ducklings is different from caring for chicks.

Raising Healthy Ducks

"All thought is a feat of association; having what's in front of you bring up something in your mind that you almost didn't know you knew"

- Robert Frost

So what do you do to ensure that the ducks grow healthy, strong, and highly productive? We will learn that in this chapter.

Encouraging the ducklings to swim

Generally, ducks love swimming hence explaining their webbed feet. They will start as soon as the day they are hatched if you let them. Do not let young ducks swim unattended. As for baby ducks, they are covered with down, which is not waterproof. Their bodies are also too fragile for them to swim at this stage of life. Ducklings usually do not produce waterproofing oil until they are about 4 weeks old hence they can be able to swim at that time. In the wild, mother ducks usually apply the oil hence they can swim immediately they hatch. Domestic ducklings hence cannot be allowed to swim before this time.

You can make a tiny swimming pool for the young ducks out of a paint roller tray. It makes such a good environment for swimming for starter ducklings. The slope in this tray creates a nice ramp that the ducklings can use to get into the pool and out of it safely.

Do not let the little ducklings swim for a long time since they might get chilled. After they are finished swimming, dry them off in a gentle manner and then place them back in the brooder to allow them to warm up. You can also allow the ducklings to sit on a heating pad that is covered with a clean towel for just a few minutes. The ducks can be allowed to swim independently after a month.

Allow adult ducks to swim without any assistance

When your ducklings become fully feathered, you can allow them to swim without your supervision since the feathers are waterproof and can prevent them from feeling chilled. Depending on the breed of duck you choose to buy the duck should have fully developed feathers in about 9-12 weeks.

Be cautious with older ducks

You should always ensure that your ducklings are supervised at all times as they are still learning to swim and growing their adult feathers especially if you are going to let the small ducks swim in an outdoor pool/pond. This is because older ducks that might be sharing the same pond or any other water source with the young ducks may try to kill or drown the young ducks.

Keep your ducklings safe from predators

Predators usually target young ducks. You can leave your ducks to roam about when they are completely adults but remember some predators may manage to get to them. You should ensure that you make efforts to ensure your ducks are safe. If you are raising your ducks in a barn outside, you should make sure that no animals can get to them. Foxes, dogs, snakes, wolves and other predators may cause harm to your ducklings if you are not careful. If you choose to raise your ducklings inside your home, you need to keep them safe from cats and dogs which may attack or even play with your ducklings roughly hence hurting them.

One you move your ducklings from the brooder to a large pen, you should ensure that there is no way that any predators can get in.

Do not get too emotionally attached to your ducklings

Sometimes it can be very tempting to cuddle the fuzzy ducklings but getting too close to the ducklings might cause them to get imprinted on you too strongly. To ensure that your ducklings grow into healthy and independent adult ducks, you should just enjoy watching them as they play with each other but avoid joining in!

The case would be different only when you intend to keep your duck as a pet. In such a case, you should handle the ducklings often in this stage. This makes them social hence forming a bond with you.

Move your ducks to a larger space

Once you ducks do not fit in the brooder, you should transfer them to any structure that has a latching door. Feed them with adult duck feed and give them time during the day to spend splashing and swimming in the pond. Just ensure that you bring them back into their shelter at night to protect them from predators.

Ensure you provide your adult ducks with plenty of water, as they feed so that they can be able to swallow their food well and clean their beak vents. Keep a bowl/dish of water near the food source for the ducks. They need to have something to drink as they eat to avoid choking.

You also do not have to provide your adult ducks with a pond. Cleaning ponds can be pretty hard at times. In fact, if left unclean, it could be a threat to the life of the ducks. You can use plastic wading pools to provide your ducks with a place to swim. This is a great choice because they are rather inexpensive, easy to move and easy to clean. To limit the amount of mud that develops underneath the pool, you should spread sawdust, sand and pea gravel underneath it. Change it once or twice per year.

Ensure your adult ducks take a balanced diet

Even though your adult ducks will often forage for grasses, slugs and other food on their own, you will need to supplement that food with some balanced and nutritious food. You should use commercial waterfowl feeds. If you are unable to find this type of food, you can go with non-medicated chicken feed or commercial game bird feed. You may also need to add on some supplements of calcium or grit to help the duck with strengthening of bone and digestion

As the adult duck continues to age, its nutritional requirements will begin to vary. Mainly, younger ducks are not supposed to receive high levels of calcium unless you are raising them for their meat.

Ensure the duck shelter is adequate

Not only ducklings fall prey to predators and victims of bad weather- adult ducks do too. For this reason, you should build a shelter that has the ability to protect the ducks from both elements. The shelter should also provide the ducks with a quiet place that they can rest peacefully. Ensure that the shelter is well ventilated and big enough to allow some space for the ducks to groom themselves in.

You can put them in a coop, insulated house or an enclosed pen. It should not be necessarily perfectly snug. If you are able to spend energy and time on it, you can consider acquiring a shepherding dog, which you can train so that it guards your ducks all day. You should also keep your ducks fenced in. Even if you allow the ducks to roam around your yard, you should also ensure you surround the duck's grazing area using a protective fence. There's no need to put up a high fence. Just 0.6 – 0.8 m (2 – 2.5 feet) is high enough. This is because most ducks will never try to jump over as long as they are cared for well. If you happen to have a flying breed of duck, you should clip their primary feathers on one wing one time each year to keep them grounded.

Mind the health of duck

Ducks are quite resistant to diseases and worms that often affect chicken, but you have to ensure that all their basic health needs are met. In order to ensure that your ducks are in good health, you should ensure that they get plenty of exercise in an open space. Ensure that the ratio of male to female ducks does not exceed 1:3 to prevent excessive stress on females. Keep an eye on all your ducks and note down any signs of illness including bloody diarrhea, lethargy, changes in water and food consumption and ruffled feathers. If any of your ducks is ill, you should immediately quarantine it to avoid spreading then treat it.

*Key point/action step

It is your duty to do everything possible to protect your ducks from harm. Although they are generally quite resistant to diseases, you will obviously need to keep the area you keep the ducks clean. Supply them with water and set up proper mechanisms to keep the ducks and ducklings protected from predators.

How to Apply What You've Learned?

By now, you have gained some knowledge on how to actually rear ducks with ease. The next thing you need to do now is to start by getting a few ducklings (probably a pair) or hatch your own ducks if you have an incubator. Follow the guidelines provided in the book to only keep the viable eggs in the incubator.

Once you have the ducklings, you should provide enough warmth, feed them, and watch out just to make sure they do not drown or hurt themselves. In about 4 weeks, your ducklings should be able to live comfortably since they will already have grown the necessary insulation mechanisms to keep off cold. The great thing about raising ducks is that you don't need to worry about them getting sick too often since they are usually very resistant to diseases, cold and pests. And if one of the ducks gets sick, ensure to quarantine it just to ensure that you don't end up putting the other ducks' lives at risk.

How To Raise Strong & Healthy Pigs

For most people, the misconception that pigs are a dirty and difficult animal to rear has led them to settle for the cheap pork sold by meat processing companies. However, what most people are unaware of is that, healthy, pigs are one of the most social animals there is. If you are reading this book, then chances are you have already decided to rear pigs at your home, and are looking for important guidelines to carry you through. This book contains a simple step by step procedure, including the breeds of pigs available, food and water recommendations, as well as how to handle pests and diseases.

Once your pigs have matured, you will have several advantages, which include their meat either as pork, bacon, sausages or ham, all of which are a good source of good quality protein.

Starting Out

"I am fond of pigs. Dogs look up to us. Cats look down on us. Pigs treat us as equals."

- Winston S. Churchill

In this age of preservatives, overcrowded feedlots, hormones, antibiotics and quick cure methods, the only practical way of being assured of quality meat is producing your very own. Moreover, raising a pig is a project particularly suited for a beginning or small farmer for three main reasons. For starters, it has a low investment capital, it is a short-term project, and you can acquire a substantial amount of quality food from your family garden for the pigs at negligible cost.

How many pigs to raise

The truth is that a lone hog does not grow well at all, as he enjoys having company. The challenge is that the average family today does not actually eat more than one pig per year. There are two possible ways out of this dilemma. You could find a friend who is interested in raising a pig but does not have a place to do it or you could consider raising a second one for your comrade, who will then meet his share of the expenses and assist with the butchering.

It will be much less of a hustle to raise two pigs as compared to one; the hogs will have each other's company, and you will subsequently be doing someone a favor! The second option is to go ahead and raise both pigs, then butcher one and sell the other. Although there are laws in all states regarding the butchering and selling of meat, in general practice, you are less likely to be bothered as a small farmer who breeds for himself and sells to his friends. If you are in doubt about the feasibility of selling your home butchered meat, ask around for neighbors and friends who would be interested in purchasing fresh pork next fall. While fresh pork is not really that pricey, you should be able to get your returns, and a little extra to make up for the one you decide to keep.

Buying pigs

Young pigs in farm communities are usually offered for sale in the newspapers in the course of spring and the summer months. If none are listed in the local papers, check out the farm supply stores or visit a stock sale. However, avoid the temptation of buying your pig with the first sign of spring fever. For starters, you have probably not planted your garden yet, which means that it is no condition to feed a pig. Moreover, if you buy too early, you will end up with a costly, overgrown pet that is too expensive to maintain. You should know three things before you buy.

1: Pigs are nurtured and sold when they are 8 weeks old. Therefore, avoid buying pigs that are weaned at 6 weeks because they will not be as healthy.

2: It is generally accepted that a pig should be butchered when it is 6 months old so, if you continue feeding him after this age, you will only be investing money into the pig that no one will give you back.

3: Unless you have a walk in cooler at your disposal, you should not butcher the pig until frosty weather pulls in. The temperature range should be between 30 degrees and 40 degrees F in order for the meat to hang and cool after being butchered.

Good breeds to raise

*The kunekune of New Zealand

Kunekune is a word used to mean fat and round. The kunekune pig is a small breed with a short snout and short legs. Some hang hanging tassels from the lower jaw. These pigs have varied colors and textures, and there is a clear distinction between their winter and summer coats. They feed on grass, and are not inclined to roam. They also have an excellent ratio of fat to meat. This pig variety is sociable, and likes the company of humans, so you and your children can handle it safely, making it a very good pet.

*American mulefoot

This breed has a characteristic solid hoof, is hardy, fattens easily, and has a gentle disposition. At two years of age, this pig weighs 400 to 600 pounds.

Berkshire or kurobuta

This is the oldest pig breed in Britain, and has its origins in the Berkshire County. It is a large breed, and is black in color, with white legs. This hog matures quickly.

Red wattle

These rare hogs originated from New Caledonia, a French island, and were brought by the French to New Orleans in the late 1700s. You will find its excellent flavor resembling that of beef in taste and texture, and the pig has a lean and tender carcass. It is also a tasseled pig that adapts well to climates and is a brilliant forager. It weighs 600 to 800 pounds, but some can reach up to 1500 pounds.

The Guinea Forest

These rare hogs are believed to have originated from the Guinea Coast in Africa and then traveled widely through the slave trade to England, Spain, France, and America. The breeds were a large and square breed with pointed ears and reddish bristly hair. These pigs are excellent foragers and hardy grazers that you can raise on pasture and still get pork and lard. Guinea hogs today are small, about 15 to 20 inches in height and 150 to 300 pounds in weight when fully grown, which makes them perfect for your small farm. They have varied colors, but are often black and hairy. They are easy to care for and are very gentle, which makes them very popular with children.

Tamworth

This is a small, red-coated, thrifty, rugged and active pig breed, whose origins are still unknown. It is however one of the oldest breeds, and is a descendant of native European pig stock and wild boars. Its ham is firm and muscular, but lacks the bulk and size found in several other breeds.

*Basque pigs

This is an endangered breed that once dominated the extreme southwestern Pyrenees. This French pork was ignored in favor of other breeds due to its high proportion of fat and low growth rate. After becoming almost extinct in the 70's, some breeders have since saved the breed in appreciation of raising their pigs in regards to ancestral traditions. The Basque pigs are a weather hardy breed that are alert, lively and great foragers in the wild for fruit, peas, acorns, chestnuts and grass. The breed is white with black spots, has extra tasty meat, but tends to grow slowly. They are good bacon and lard producers because of their ability to deposit fat easily.

*Cinta senese pigs

This is a very rare breed from the Tuscan native swine family, and is the only one that survived extinction. It is very resistant to bad weather, and is an almost savage breed. Therefore, it can be a secure food reserve for you and your children. However, this swine grows very slowly, which is why most farmers neglect them in favor of other breeds that grow faster. The cinta senese pigs are bred half-wild, eating roots, digging in the dirt and feeding on mushrooms. You can butcher the pig at two years of age. You may use its low fat, fragrant pork for cooking, but its main use is in the production of different types of tasty cold cuts.

Iberico pigs

The origin of this black breed of pigs can be traced back to ancient times, in Spain, Portugal and the southern and central territory of the Iberian Peninsula. Since these animals live freely, they are constantly on the move, which makes them burn more calories as compared to other species. The pig builds up fat between its muscular fibers and under its skin, which is the reason for the typical white streaks that are characteristic of the ham.

Meishan pig

This is a small to medium sized breed from the Taihu breed. It has black, wrinkled skin and large drooping ears. The breed is native to Southern china, and is most famous for its large litters of fifteen to sixteen piglets. These are perhaps one of the most prolific pig breeds in the world, and are known to produce two litters in a year. The USDA Agricultural Research Service to the United States imported this particular pig breed because of its fecundity. However, it lost flavor because of abundance of fat and its slow growth rate, although it matures early. While they are fat and slow growing, they do have a very good taste. They are also said to be resistant to certain diseases.

Large black

This breed originated from Chinese and found its way to England in the late 1800's. It is known for its taste, general hardiness, and pasture foraging skills. Large Blacks also feature short black hair, a long body and wide shoulders. When you butcher the pig, even at two hundred pounds, the excellent bellies, short muscle fibers, and micro marbling all create moist meat and exceptional bacon with a tasty flavor. The color of its coat makes it resistant to several sun borne conditions and its grazing ability and hardiness make it a sufficient meat producer. These pigs are very docile in nature and tend to move more deliberately and slowly than other breeds. The Large Blacks are also slow maturing and are known for their prolificacy, milk capacity and mothering ability. The sows bear 8 to 10 piglets.

*Ossabaw Island, Georgia pig

This endangered breed originated from Spain and lives off the coast of Georgia. These pigs have a long snout and heavy coat. Their popularity within the chef community has been boosted by their fat quality and marbling. They are small and weigh less than 200 pounds and are less than two hundred inches tall at maturity, although they grow much larger in captivity. Its meat is considered a heritage product that is specially suited for use in whole pig roasts, cured meats, and in pork.

*Key point/action step

When looking for your own pigs, ask for either barrows (the castrated males), or sows (the females), because meat from an uncastrated male (boar), has a very unpleasant taste and odor. If possible, go for the huskiest looking hogs of the litter; the plump-looking hams that have short legs. You will need a box or wooden crate to transport them in since a pig does not handle like other animals. Ensure that you choose a suitable breed that you will enjoy having rather than one that will disappoint you.

Housing Your Pigs

"Quality, affordable housing is a key element of a strong and secure Iowa."

- Thomas Vilsack

Once you have determined the kinds of pigs you want to keep, the next step is to design a house to shelter your animals. The best way to house your pigs is to provide a shelter that can keep them comfortable in different seasons. The ideal ones are built from either bricks or concrete to make it easier to clean. However, you can still make your shelter using wood, and then surround it with a wire mesh. The work of the mesh is to prevent the pigs from nibbling the wood.

In addition to the structure, it is also important for the pigs to have access to shallow ponds or mud. This will come in handy during hot days, when your pigs need to cool off. You must also ensure that food and water are readily available when raising swines. However, since pigs are regarded as strong creatures, you need to secure them in strong enclosures at all times. The optimum length of a typical pig house is roughly 4 feet tall. The average weight of a typical swine is roughly 150 pounds, but some can weigh up to 800 pounds. Therefore, 8x6 ft area for each pig should be sufficient enough to allow them to move around the shelter with ease.

When constructing a pig house, it is crucial that you get a good guide and execute your ideal shelter very well in order to avoid extra expenses caused by repairs. The materials needed are usually concrete or bricks, with wire mesh, six metal posts, straw, plywood, and hog panels. The roof is normally made from wood. To achieve good ventilation, raise the roof to 12 inches above the wall. For adequate bedding, you may want to place straws all over the floor. Moreover, it is also advisable to plan the location wisely. It should be within an area that has enough water supplies such as an outdoor well or a pond. Some farmers find it useful to dig about a foot or so into the ground before building the house. This is especially beneficial in cooling and insulating the inside of the pig house in order to keep them comfortable all year round. Generally, it is not that difficult to obtain a pig shelter, as long as your objective is to provide a comfortable dwelling for your pigs.

Shelter requirements

The temperature range needed to achieve maximum pig productivity is known as the thermoneutral zone. The heat production of pigs within this zone does not depend on air temperature, and is therefore determined by its food intake and live weight. Critical temperatures will vary depending on the specific conditions in your piggery and your pigs' total weight. However, if they spend more time shivering or huddling, and eat more than normal, they are usually cold. On the other hand, if they avoid body contact with their counterpart mates, foul clean areas of their pens, eat less and pant at more than fifty breaths per minute, then they will become warm. The highest tolerable thermoneutral zone temperature is between 6 to 8 degrees Celsius, beyond which serious problems are likely to set in.

If the temperature in the immediate surroundings of your pigs falls below the lower critical temperature, which is the lowest tolerable temperature, your pig will be inclined to maintain its body heat using some of its energy. Older pigs can withstand lower temperatures for short periods without signs of obvious ill health, at the expense of the efficiency of their food conversion. It is also important to note that the LCT decreases with the age of the pig. The ideal temperature for newborn piglets is usually between 27 and 35 degrees Celsius. However, their ability to withstand cold temperature becomes limited, as the piglets grow older. You are more likely to lose some piglets very quickly if the microclimate stays below 16 degrees Celsius. Fatal chilling will take place within minutes at temperatures below 2 degrees Celsius, unless you provide them with warmth.

However, if the area is not draughty, your pigs can withstand low temperature. Some areas to keep an eye on are open-ended trenches that allow draught through the slats, wall cracks or near floor level, as well as uncovered heat lamps in cold buildings that can result in a draught when hot air is displaced by cold air at the floor level. In addition, be sure to use creep boxes or covers to reduce draughts and retain warmth.

It is usually quite easy to warm a dry concrete floor for your pigs. However, while concrete tends to retain heat considerably well, it also increases the harmful low temperature effects when damp. Your pig will pass considerable heat into the damp concrete floors, despite the air temperature being reasonable.

Untreated wood shavings or dry straws are excellent insulations for very young pigs against cold conditions. If the temperature in the immediate surroundings of your pigs' shelter goes beyond the UCT, it will make the pig severely distressed. As the pig ages, the UCT reduces. Young pigs are the most susceptible to cold, while larger and older animals are more vulnerable to rising temperatures. In fact, temperatures above 27 degrees C are widely considered undesirable for breeders, finishers, and growers. However, if your pigs have sufficient air movement in the shelter, you can reduce heat stress in dry climates through spray or dip cooling. The resulting water evaporation from the skin of the pigs can significantly reduce excessive body heat.

Regardless of environmental conditions, you need to provide a minimum amount of fresh air to your pig house in order to eliminate odors, bacteria, airborne dust, ammonia, carbon dioxide, and water vapor. However, ventilation tends to reduce the temperature in the shelter, which is why it is important to insulate the walls and roof to minimize heat gain or loss through conduction, and keep it draught proof to limit uncontrolled air change. If your insulation does not create its own vapor barrier, you can protect it with a vapor barrier to reduce condensation within the shed. This serves to protect the interior linings, as well as reducing the amount of ventilation you will need to prevent condensation. Be sure to direct cold ventilating air such that it creates air circulation without flowing directly onto the animals within the pig house. A conventionally and naturally ventilated pig house normally involves using side wall vents together with a ridge vent.

Building orientation

If the long axis spreads from east to west, long and narrow buildings tend to be warmer in winter and cooler in summers. You should situate your pig shed to make the most of prevailing winds for coolness in the summer months. Conversely, you should protect ventilation openings in winter against prevailing winds. You can achieve this by planting trees in a shelterbelt that will not impede with the airflow needed for cooling during summer. These will soften the visual impact and enhance the appearance of your piggery.

*Key point/action step

Maintaining adequate temperature in the pig's shed is very important for the survival of your pigs. Your goal is to ensure that the shed is not too hot by providing adequate ventilation while also making sure that it is not too cold by providing the necessary insulation. Also, have in mind that the age of the pig determines the kind of house they need. While a large pig is susceptible to high temperatures, a young one is more susceptible to cold temperatures so have this in mind when building a pig shed.

Feeding Your Pigs

"Animals feed; man eats; only a man of wit knows how to eat."

- Jean-Anthelme Brillat-Savarin

Pigs are omnivorous, just like us, and can eat a wide range of foods. They also need a balanced diet like us including fiber, proteins, energy, vitamins, and minerals to grow strong and healthy. In keeping pigs, feed will be your biggest expenditure, so it is useful to get it right. Pigs require a constant supply of fresh drinking water, like all the other animals. Pigs tend to tip the trough to make a wallow, and are known to stand in the water trough and wash themselves in it, so you have to check, clean and refill the trough on a regular basis.

Galvanized troughs are strong and easy to clean, as well as hard for the swine to tip over. You can also get automatic drinkers, but these are not as much enjoyable for the pigs. If you have the right equipment and knowledge, you can mix your own food, but most people prefer using commercial pig food. There are several different feed producers, including both GM feeds and organic feeds. Be sure to find a dry and rodent free place to store the bags. Store the food in a rodent proof container once you open the bags. In addition, make sure you sweep out and get rid of any spillages. Check regularly to ensure that the bags have not been damaged by rodents, and if they have, deal with the damage right away. Always check the expiry date before purchasing your feed. It may retain the oil, fiber and protein value, but its mineral and vitamin level will reduce after it.

Pigs are usually fed twice per day. However, the amount of feed will depend on the reproductive state and age of your pigs. A foraging pig will get some of its food from natural sources, if the foraging area can be able to provide it. This would include apples, acorns, brambles, grass and, yes, earthworms. It is crucial that you supplement these with an all-rounded compound feed so that your pigs receive all the essential nutrients they need. However, foods such as carrots and potatoes do not cater for the nutritious value of compound feeds, and should therefore not be fed to the pigs. In fact, it is illegal to feed pigs with any household waste. You can, however, feed them with vegetables and fruits from non catering premises.

Pigs like to eat their feed when wet, so it is advisable to add surplus goat's milk or water to their feed, so long as the milk has not passed through the kitchen, after which it would be considered catering waste. In addition, unless you are registered with Animal Health, ensure that your pigs' diet does not consist of more than 80 percent of waste milk. Pig troughs are especially useful for sheds with more than one pig, as this will ensure that even the more timid members get enough to eat, or alternatively spread the feed around the ground widely. The latter should be done only on clean areas, but will inevitably lead to more feed waste.

Gilts

The gilts (young sows), will require 5 ½ lb (2.5 kg) of meal, cakes or sow breeders pencil per day. You should keep these until right before farrowing. However, be careful, keeping in mind that a maiden gilt is still in its growing stages, and needs to feed her unborn child. This means that she needs sufficient feed to be able to perform both functions. You should increase the rations of the gilt gradually to 4 kg after three weeks. This will help maintain her best. During gestation and after service, you may reduce the feed to 5 lb (2.4 kg) per day.

After farrowing

As soon as the pig has given birth to her litter, you must get her enough food for her to stay healthy and to provide sufficient quality milk for her piglets. Be sure to feed the sow roughly 6 lb or 3kg of meal, cakes or sow breeders pencils per day. If the sow has given birth to more than 6 piglets, you should feed her with an additional 0.5kg per each extra piglet. You can reduce this to 1.5 to 2 kg after weaning.

*Key point/action step

Proper feeding of the pig will determine if you will have healthy pigs or not. Remember that pigs also need clean water too so keep checking on the water in the trough as pigs are known to tip the trough and even step on the trough. If you have a pig with piglets, ensure that you provide more food so that the pig can provide adequate and quality milk for the piglets.

Hygiene

"Civilization is the distance that man has placed between himself and his own excreta."

- Brian W. Aldiss

As a pig farmer, you need to take precautionary measures to protect the health of your animals. Maintaining good health is not only vital to ensure acceptable standards of animal welfare, but also to maximize on their productivity.

External measures to prevent pathogens

*Your pig farm should be located far enough from any other farms, especially in the direction of the dominant wind. Flies can spread some infectious particles over short distances. Rodents too tend to spread diseases from one farm to the next.

*Purchasing policy: Purchasing should be limited and selected carefully on vaccination and sanitary status, when it comes to a breeding farm. Minimize purchasing sources in fattening farms.

*Quarantine: Once you bring in new animals into the farm, you should keep them in quarantine for veterinary observation. In fact, this is a legal obligation in several European countries. The practical period is usually 4 to 6 weeks.

*Minimize visitors. Make sure that anyone entering the farm is wearing boots and an overall. They should also wash and disinfect their hands before and after leaving the shed.

*Vehicles: In order to prevent cross-contamination, vehicles transporting animals are required to be cleaned thoroughly and disinfected after every trip. The golden rule is to clean first and then disinfect. A slightly alkaline detergent should be used for cleaning, as this will remove organic dirt such as fat and various proteins. Acidic products remove inorganic dirt such as lime scale. However, an excessively alkaline solution or one that contains chlorine or sodium hydroxide will corrode the truck, especially the parts made of aluminum. It is not possible to remove these specific kinds of dirt by using high pressure water alone. If possible, ensure that tracks that are not disinfected do not enter your farm for pick up, if you are raising several pigs for sale.

*Footbaths: Install a footbath at the entrance of every house. Make sure that the boots are properly cleaned before dipping them into the foot dip. In addition, the disinfectant should be powerful enough to kill all microorganisms. Ideally, you should renew it daily.

Internal measures to deal with pathogens within the farm

*Cleaning and disinfection: It is crucial to employ cleaning, disinfection and sanitary measures after every cycle to prevent development and progression of pathogens within your farm.

*Key point/action step

You want healthy pigs not only for the good of the pigs but also for your good and for that of your family. You would not want to have sickly pigs that can easily transfer pathogens to the home when you are in contact with them. Therefore, ensure that you emphasize on cleanliness.

How To Recognize Disease In Your Pig Farm

"He who cures a disease may be the skillfullest, but he that prevents it is the safest physician."

- Thomas Fuller

The first priority when it comes to managing disease in your pig farm is early recognition through using the senses of smell, touch, sound and sight to detect the sick animal and to distinguish it from the healthy animals. If possible, it is advisable to carry out a clinical examination of all your pigs every day.

The use of sight

*Lack of appetite is usually normal when an animal is bred in isolation, such as a confined sow. However, this is relatively hard to detect in group housed animals. If you notice a drop in feed intake or failure to eat by otherwise normal pigs, be alert at once and check for availability of water. Loss of appetite in all pigs in a group could highly be due to lack of water. Look for signs of disease if the water supply is okay.

*A dull appearance of listlessness is also an early sign of illness.

*If you notice rising of hair over one of your pig's body or shivering, take action at once as this is a common symptom of disease, and is an early sign of joint infections or streptococcal meningitis. If a pig is shivering with its hair on end as it lies on its belly, it could mean that the animal is either lame or scoured from a generalized septicemia, which is a bacteria in the bloodstream.

*Loss of weight in one of your pigs is a first sign of dehydration or loss of appetite due to pneumonia or diarrhea.

*Discharges from the eyes or nose mean that your pig has an upper respiratory infection. If it has excess salivation from its mouth, it could indicate an exotic disease like vesicular disease. A discharge from the vulva of sows could indicate endometritis, pyelonephritis, cystitis, or vaginitis.

*There is a wide range of diseases that could cause fecal changes, but sloppy feces could actually be normal. Watch out for signs of blood or mucus indicative of salmonella infections, swine dysentery, proliferative haemorrhagic enteropathy or gastric ulceration.

*If you notice vomiting from one of your pigs, it could be a sign of such diseases as transmissible gastro enteritis, or gastric ulceration in individual pigs.

*Skin changes can help you identify diseases, typified by chronic or acute lesions of lice and mange, although the former are now quite uncommon. Extreme bluing could indicate acute bacterial septicemia, acute viral infections, or a toxic condition, such as seen in mastitis, PRRS infections and flu.

*Respiration rates: If you have identified any of the mentioned changes, observe the rest of the pigs and compare the rate of respiration of both the suspect pigs and the normal ones. Determine if the breathing is deep with considerable chest movement because of consolidation of the lungs, and scarcity of oxygen, or if it is a very shallow breathing indicating pain and pleurisy.

Observing the group

You should set daily and regular time to examine all your pigs. You should also assess the environment of the shed by noting the following:

*Humidity

*Smell

*Ammonia levels as seen through breathing and the effect on eyes

*Temperature

*Ventilation

*Pig behavior

*Appetite

*Human reaction

*Abnormal changes in slurry and bedding

Changes in behavior

Pigs are known to be social animals that prefer being part of a group in a healthy condition. However, when ill, your pig will tend to rest on its own, and can even be isolated by the other members to the extent of being attacked. You should regard abnormal lying patterns with suspicion. Huddling, on the other hand, is common where several pigs are sick or if the environment is inadequate. Be alert if you notice reluctance in your pigs to show interest or rise in the presence of an observer.

***Key point/action step**

No one knows your pigs better than you do. Therefore, ensure that you note any change of behavior as this could mean that they are unwell or are uncomfortable.

How to Apply What You've Learned?

Consumers have for long been happy to purchase low quality pork without giving much thought as to how it was produced. In truth, most of us have been unaware of the changes in the housing done at the pig industry. However, you may have noticed that the quality of the pork had changed and that it had become bland and dry. The use of words such as "quality assured", "all natural", and "grain fed" have all been served to keep you in the dark just that bit longer. When all is said and done, in order to raise strong and healthy pigs, you will need good management skills to produce a desirable product, and follow at least these five principles:

*Space: Sufficient room for the pig to escape any confrontation and behave naturally

*Diet: A well balanced diet with all the nutrients essential for your class of pig

*Water: A ready supply of fresh and clean water, or else your pigs will refuse to eat

*Shelter: Appropriate space with enough protection from environmental elements

*Stockmanship: Pigs that are handled well are happy pigs!

How To Raise Strong & Healthy Chickens

With the current economic times coupled with the increasing number of diet related health problems, the best you can do is to have your own supply of your favorite foods. Whether it is growing fruits and veggies or raising some animals like poultry, you will definitely be a lot better placed than counting on the GMO chicken that you buy from the local meat market.

If you are thinking that it is impossible to rear chicken even in the tiniest of places, then think again. Rearing chicken is a lot easier than you have been thinking if you've never tried it. If you don't know what to do, then this book will offer all the help you need. It will help you to decide whether to keep layers or broilers, prepare the chicken coop, bring the chicks home, feed them and protect them from environmental hazards and diseases and feed them until they are fully grown and ready to lay eggs or ready for slaughter.

It doesn't matter which stage you are in rearing chicken. Whether you are just contemplating about buying the chicks or are thinking of how to deal with the diseases that affect the chicken in their growth phase, you will find this book helpful.

Beginning At The Beginning: Choosing The Type of Chicken to Raise

"Beginnings often spell the end. Thus, handle these as if they were the most important things in all of the earth"

-L. Ron Hubbard.

Rearing chicken is a process. The first thing you need to do is to decide what kind of chicken you want to keep; is it broilers or layers. So, how can you choose between these?

It is important to begin at the beginning: an egg laying poultry is referred to as an egger or layer whereas the broiler poultry is one reared primarily for meat. Thus, a layer should be raised to produce many large eggs without necessarily growing too much. On the other hand, you will need to raise a broiler to yield a lot of meat and hence, be able to grow well.

Good management practices will be essential for poultry production, regardless of whether you choose to go the broiler way or the layer way. The thing to understand here is that the management procedures differ for both layers and broilers. These will include temperature maintenance as well as hygienic conditions in both housing and poultry feed as well as keeping pests and diseases at bay. Let's take a quick overview of how these two are different in the way you raise them.

Broiler chicken

The nutritional, housing, and environmental requirements of broilers differ from those of layers. Broiler feed should be rich in vitamins to boost the growth rate and superior feed efficiency. The ration for broilers will also be rich in proteins as well as have adequate fat. Absolute care is to be taken to minimize mortality as well as maintain feathering as well as the quality of the carcass.

Layer chicken

These show 2 distinct phases in their life; the growing period and the laying period. During the growing period, the issue of space is vital: they need enough of it. If you overcrowd them, the result is suppression in growth. However when it comes to feed, it should be administered in a restricted and calculated manner.

When the laying period comes along, both adequate space and ample lighting are of great importance. Feed your chicken with minerals, vitamins, and micronutrients to influence the laying of eggs. You may use agricultural by-products, which will be cheaper yet more fibrous and beneficial as a result.

So, if you've settled for broiler or layers, you should start preparing where you will house them and provide ideal conditions for them to guarantee maximum growth and maximum productivity while ensuring minimal mortality.

Key Point

You need to settle on what it is you want most. Is it meat or is it eggs? You can as well settle for some mixed breeds that can supply eggs and meat for your family. But these will probably have different rates of growth; their growth is likely to be slower than that of the pure breeds.

Preparing A Home For Your Chicks

"Give me six hours to chop down a tree and I will spend the first four sharpening the axe."

—Abraham Lincoln

Step I: What to do before the chicks arrive

Usually, this is mostly about setting up the brooder. It will make little sense to bring your load of chicks over and not have a home ready for them. This is a strong prerequisite to raising strong, healthy chicken. The brooder will need to provide adequate ventilation, protection, and warmth.

What to do

Make sure there is a ¾ square foot space per chick

Cover the brooder floor with about 4 inches of litter

Place a heat lamp above the brooder. For the 1st week, you will need to keep brooder temperatures at 95 degrees.

Getting your chicks settled

Regardless of where the chicks are from (feed mill, hatchery etc.), you need to put your chicks in a brooder right away.

Show your chicks the location of their water. You will do this by gently dipping their beaks in the water one by one.

Watch the chicks for a while to make sure they are neither too cold nor too hot.

Tip: Pasty butt: This occurs when poop is stuck on the chick's downy feathers, resulting in demise, as the chick cannot poop. When this happens, clean the refuse off with warm water and put a little bit of olive oil all around the vent.

Step II: Raising the broiler and layer chicks to strong and healthy chicken

This book emphasizes on the term "raise". Here is the thing: to raise, especially with regard to chicken, means to take care of them right from the time when they are tiny chicks with tiny pairs of legs. If you do the starting duties right, you increase the chances of having healthy chicken immeasurably. Here is what to do to ensure your chicks grow up to be healthy, strong chicken:

Brooder

The first thing to do is to confine the chicken to a brooder that has solid sides each about 18 inches high so as to keep drafts out. It is important to make sure that the brooder is located near a heat source, most preferably a heat lamp. For each chick, allocate about 6 square inches of floor space. Finally, the chicken need to be as protected from predators as possible. Locate the brooder in a safe and secure place.

Brooder Floor

Pine shavings are the best material you can use to cover the floor with. If pine shavings are a bit rare where you come from, any other absorbent bedding material will do the job. Steer clear from using kitty litter or cedar shavings. Stay away from using newspaper as well. For the duration of the first two days, cover the littler with several paper towels/tissue paper or even a piece of aged cloth so as to keep the chicks from eating the litter until they can locate their food. After this period however, you can remove the paper towels or pieces of cloth.

Temperature

Ensure to turn on the heat lamps at a day or more before you can bring the chicks home. This should make sure that the floor, walls, air, and bedding start warming up and even any disinfectant you've used has dried. You will also ensure that the air temperature is ideal.

For the 1st week, the chicks require a steady temperature of about 95 degrees Fahrenheit. This should be kept steady at all times. For every week that passes, drop the temperature by about 5 degrees until the surrounding room temperatures outside the brooder is equal to the temperature inside the brooder. However, drop the temperature no more once you hit the 60 degrees Fahrenheit mark.

Feed

For all chicks ensure to use starter- we will talk about this in detail later in the book. For the 1st two days, sprinkle the feed on a white paper plate or even some paper towels of the same color to as to make it easier to find. You should make sure to have feed available at all times in the feed dishes that you allocate the chicks.

Water

With baby chicks, you need to put their water in a container that is both shallow and narrow. The reason for this is to cut off any chances the chicks may have of drowning. Dip their beaks into the water containers gently as you place them into their brooder so that they may develop a memory of where it is. At all times, always have water available.

Note: Ensure to fill the chick's waterers with warm water then dissolve a tablespoon of molasses to boiling water to let it dissolve then add in some more water to make up to 4 liters or a gallon of lukewarm water. Do this before your chicks arrive. This should be the first thing you give to the chicks when they arrive (give them for 3 days) to facilitate recovery from the traveling. The chicks can survive for about 3 days by relying on the yoke sack that they draw into their body right before they are able to hatch but they need to drink water- use the strategy mentioned above to give them water.

Handling

It is a mistake to handle baby chicks too much. The reason for this is that handling them to a high degree only serves to stress them and in turn, makes them not grow well. Handling them too much may also kill the chicken eventually.

Key point

When chicks are contented, they usually are fairly quiet, often spread out over the brooder and either eating, drinking or eating. If you realize the chicken are making way too much noise, there is a high chance that something is wrong. Usually, the reason is that the chicks are too cold. If you notice them spread out against the walls of the brooder and panting heavily, then they are too hot for comfort.

Housing: The Chicken Coop & Chicken Run

"Once your mind is made up that chicken are for you, it is only logical that the very next thought to cross your mind should have everything to do with housing."

-Tim Daniels.

The Chicken Coop

As far as expenses go, nothing will be as expensive as a chicken coop for your chicken. But if your chicken are to truly grow up as healthy as can be, you really do have to invest in a good chicken coop. The good part about all this is that there are multiple designs and types out there in the market. As such, you will be able to come out of it feeling like you indeed have gotten value for your money.

Usually, it is a case of "you get what you shell out for" with wood but then again, there are lots of farmers who have been able to keep their cheap wooden chicken coops in superb conditions for periods of up to 7 years by simply employing the services of a regular coat of wood preserver. Regular maintenance (for instance, replacing any latches that have rusted and roof felts that the winds have picked up) does not hurt. Once you make up your mind on raising chicken and set out looking for a coop, then do take a very keen look on the overall workmanship and the quality/thickness of the wood used as this will be the number one indicator on whether you should be paying a higher price or not. Very obviously, the better the wood is, the longer your chicken coop will last. However, the wood will also be a lot more expensive.

Chicken houses, like most other things in life, can be either immensely beautiful or very basic. If anything, you could set out with a hammer and nails and build your own coop out of cheap pallet wood; as long as it does its job (which is to keep the inner conditions dry and draft free), everything else matters little.

Wood for your chicken coop, ideally, should always be moisture treated. This is to keep off the rot in the first year. The screws, fittings, and nails need to be galvanized to stave off the rust.

Here are a few things you must do to ensure the very best coop for the healthiest of chicken:

When you first buy a chicken coop, apply a coat of paint to give it a longer lease of life.

If you are treating your chicken coop, check to see if the product you are using is animal friendly. Leave it out to dry before you allow your birds into the coop. If you are unsure, call the numbers provided on the back of the tin and ask for advice.

Essential features of the house are nest boxes, ventilation (which should be adjustable) and roosting perches. However, you must consider the ease of collecting eggs as well as the ease of cleaning.

The other thing to keep in mind is that spending that little bit more on a chicken house that comes with a droppings board will save you hours of struggling in the long run.

The bare minimum, when it comes to chicken, is to have dry conditions, away from draughts, safe from any predatory animals and have a private place to lay eggs. Ask yourself if your coop provides all the above. If it doesn't, get back to the drawing board.

The Chicken Run

To remain healthy, your chicken will need to step outside. Your chicken run could be in the back garden or perhaps a pen or a small fenced area. Whatever it is that you settle on, the advice here is to have your run as big as possible so that your chicken can have as much free ranging space as is possible. There are multiple benefits to chicken in set ups that offer free ranging conditions, among them happier, healthier chicken that tend to cost you less to raise and even produce more eggs.

Keep the chicken run secure

The run should successfully keep the fox out and the chicken in. If it is possible, purchase wire netting that is at least 20cm into the ground to keep off predators that might try to dig under the fence. Cover the roof of the chicken run as well.

Where foxes and other chicken predators are a serious problem and the run in question is particularly large, you may employ the services of an electric fence. The ideal set up will have 3 strands of electric wire looped back and forth in an organized manner. The first strand, kept at low level, will stop the foxes from digging underneath the run. The other 2 strands will successfully keep out the predator from climbing over the fence.

Key point

For ideal conditions, keep the floor covered with wood shavings, straw or chopped cardboard. If anything, this material will get into your veggies patch far faster than anything else. The larger the pen, the easier it will be to infiltrate. Firm up security measures the more your pen size grows.

Feeding Ideals For Strong And Healthy Layers: Ideals On What, How And When To Feed Your Chicken At Different Ages And Stages As Well As What Their Feed Entails

"The spirit cannot endure the body when overfed, but, if underfed, the body cannot endure the spirit."

-St Frances de Sales

Just as the case is with human beings, chicken at differing stages of development will require different formulations of feed. The commercially prepared rations pack quite a punch in as much as offering a balanced diet is involved. This is the leading reason why you should strongly consider commercially processed feed. The other thing to keep in mind is that dabbling in assembling feed that is home processed is not recommended. If the calculations you put in place are imprecise, it is very easy to affect the growth in young chicken as well as egg production in older chicken.

Starter Feed (Chicks: Day 1 to 8 weeks)

Chicks that are day old through to 8 weeks will require starter feed that will contain 20% protein. You should know that starter feed will contain the highest amount of protein, going by percentages; a layer will ever take in its life. All this makes tremendous sense however, given the astronomical growth rate in the first couple of months of life.

You can have starter feed in two varieties: medicated and unmedicated varieties. The medicated feed will contain amprolium (which will crop up in the later chapters dealing with chicken diseases) that plays the role of keeping away illnesses like coccidiosis. If your chicks have been vaccinated, keep them away from medicated feed, as the amprolium will combat the vaccine rendering it useless. Medicated feed is unnecessary too if the living conditions of the chicks are kept clean to a high standard.

With regard to treats, you may stop obsessing right this instant on what age is appropriate to start including these in the diet. No age is ever appropriate, as treats take up a large nutritional part that would otherwise have been offered by normal feed. Treats must also be served along with grit, which is very useful for digestive purposes.

Grower Feed ("Teenage generation: 8-18 weeks of age)

With the vast amount of proteins it contains, starter feed may well launch the body of a young pullet into egg laying before it is even ready for it. This is where grower ration comes in: with its 16-18% protein content, it is significantly less compared to starter feed.

Keep off from feeding your chicks with layer feed until they are old enough for it (roughly 18 weeks) or have begun laying eggs. This is because it comes with calcium, which may well damage the chicken' kidneys, lead to kidney stones and greatly shorten the chicken' lifespan.

If you are providing treats along with grower feed to your chicken, always accompany grit with it, especially if the chicken you are raising are not allowed to wander around and forage.

Layer Feed (The big lot: 18 weeks and above)

Layer feed comes available in mash, pellet or crumble forms. Mash is the tiniest form while pellets are the largest, with crumble somewhere in between.

Layer feed comes with around 16 to 18 percent of protein along with additional calcium, necessary for the formation of eggshells. Hens that are already laying may be fed with layer ration as early as 18 weeks or even as late as it takes for them to produce their 1st egg. If the bird is younger than 18 weeks, it is wise to keep off the feed.

While layer feed will contain calcium, a source of additional calcium, such as oyster that is crushed ought to be made available to hens that are laying in a dish that is separate. Keep this away from the feed at all costs. Laying hens have varying needs of calcium and will take in as much as they need. Never add oyster shells directly to the feed, since the excess calcium may well be harmful to your chicken.

Cut the treats away

Commercial layer feed will provide every daily nutritional element that a chicken requires. Providing the chicken with snacks, table scraps and treats in addition to the regular feed will only serve to interfere with the chicken balanced diet, at least to a degree. Limiting the snacks, even those that may be categorized as healthy (homemade flock block, mealworms and pumpkin seeds) will ensure that your flock is getting all it needs. This will also help them swerve away from issues of obesity, egg binding, feather picking and a marked reduction in egg production.

Scratch

Lumping chicken scratch in with the feed category is misguided. Chicken scratch does not qualify as feed. Scratch content will vary from region to region but generally, it consists of cracked corn and an assortment of other grains. Scratch acts as a source of energy (thinking about carbs will help you understand this). However, scratch is a poor source of vitamins, proteins, and minerals. Feed your chicken scratch in a sparing form.

In cold weather, you will expect your chicken to expend more energy than is usual. Adding in a good amount of scratch just before dusk will be great in aiding them supplement energy. Too much scratch will bring about obesity as well as obesity related fatalities. The extra calories that come with scratch will also be beneficial to chicken that are brooding.

Feeding different age groups together

Chicken of differing age groups often occupy a similar living space at any given time. This raises the question of just how to feed each of them adequately. While this situation is hardly ideal, it isn't that unique either to be honest. Often, the best solution when dealing with a group of mixed flock is to feed them un-medicated starter or grower feed with calcium available in a separate dish. The additional protein in the starter or grower feed will not hurt those birds that are older but the calcium that is contained in layer feed may do damage to the kidneys of those birds that are growing.

Key Point

You should provide age appropriate food to your chicken otherwise you risk hurting them or even disrupting their productivity as you try to help. This is especially because chicken (layers) of different age groups will require nutrients in different amounts at different ages. For instance, while they may require more protein during the first 8 weeks to facilitate growth, they will need more calcium and other nutrients when laying.

Feeding Ideals For Strong And Healthy Broilers: Ideals On What, How And When To Feed Your Chicken At Different Ages And Stages As Well As What Their Feed Entails

"Tell me what you eat, and I will tell you who you are."

-Brillat-Savarin

Poultry feeds are usually rich in minerals, vitamins, energy, proteins, and any nutrients that the chicken might need for the health of the birds, egg production, and proper growth. This means that when you add in other food whether by substituting or combination, you are in essence upsetting the balance of nutrients in the food. This means feeding any supplement or grain isn't recommended.

Note: Broiler feeding plan can sometimes vary depending on the breed and your style. You can feed them with three different kinds of feed at different times.

Immediately the chicks hatch, feed them a "starter" until they reach about 6 to 8 weeks. The starter is usually high in protein making it very effective for fueling growth. But as the chick matures, it will start needing lesser percentage of dietary protein and start needing higher level of energy.

Note: If you are going for the 3-stage process, feed the chicken the "starter" when they are 1 day to about 3 weeks old. After that, start introducing the "grower" starting from week 4 then feed that until the chicken are about 6 weeks old.

When the chicks are about 6-8 weeks, you should ensure to feed them "finisher" diet or what is referred to as "developer" diet (to pullets or cockerels, which are kept for breeding purposes). You should feed them the finisher diet until they are ready for slaughter. You should feed cockerels and pullets the developer until they get to about 20 weeks.

The finisher is usually lower in protein and very high in energy. You can use it to enhance the fat content in the meat so if you want your broilers to have plenty of fat, ensure to use the finisher. However, if you want a leaner chicken, you can continue feeding them with the grower from week 3 until they are ready for slaughter.

Note: Chicken that are kept for egg production are usually fed on pullet-type diets and not broiler diets even if they are from egg-type or broiler type stock.

As you transition from one feed to another, you must ensure to introduce it slowly. To do this, ensure to increase the amount of feed in a particular feeder over a period of about a week.

Key point

Feed broilers on Starter between 1 day and 3 weeks, Grower between 3 and 6 weeks and finisher between 6 weeks and slaughter. Finisher is specially designed to supply energy and increase the amount of fat on the chicken so you can avoid it if you don't want too much fat.

Safety & Health: Focus On The Most Common Yet Lethal Of Illnesses, Their Symptoms, Prevention And Their Treatment

"They say that prevention is better than cure, which makes tremendous sense. At times though, you will have to use cure as prevention for something worse; something like death or stunted features. Thus in essence, both prevention and cure are the same thing albeit with differing trajectories of application"

-Anonymous

As far as health goes, it is obvious that your chicken will get nowhere near it if diseases consistently plague them. It is remarkable just how poultry and humans are in their life patterns when it comes to diseases. Some diseases are more common than others; some are more lethal than the rest of them. Finally, there are those that are both common and lethal at the same time; a terrible marriage of traits that is detrimental to chicken. This chapter focuses on the latter kind and how to boost the chances of your chicken being both strong and healthy by both putting in prevention measures and treating them when those measures are breached.

Vitamin A deficiency

Vitamin A, by default, is required for the well-being of chicken as well as the proper function of their mucous producing glands. Deficiency of Vitamin A usually is because of a lack of vitamins in the diet/feed.

Symptoms

Crusty material is observed in both the eyelids and the nostrils. If it is left untreated, the accumulation of cheesy material is observed. It will often mimic respiratory diseases in its initial stages and your chicken, due to the similar damage it brings along with it, will have a lot of trouble swallowing. A severe drop in both egg production and weight is observed.

Prevention

Free-range chicken will get their fare share of Vitamin A from feeding on leafy greens. This is not the case with enclosed chicken, which will require it in their feed. Adding in a few leafy greens to the diet if you are rearing chicken in an enclosure will certainly not hurt.

Treatment

This one is as basic and direct as remedies come. Change the chicken feed and proceed to supplement it with 2 to 4 times the normal level for a duration of a fortnight. You may use a water-soluble supplement, which is available with considerable ease.

Coccidiosis

Symptoms

Your chicken will exhibit considerable weight loss, relentless bouts of diarrhea, and a loss in pigmentation. When the infection is allowed to sit so that it becomes quite severe, blood will be observed in the diarrhea and if attention is not given, fatality is the next stage.

Prevention

The way to prevent this terrible illness is absolutely reliant on drainage systems. Improve the drainage and rotate both the pen and water locations to cut down on the risk of infection in your chicken. Changing the topsoil yearly in a floor pen will reduce the risk of coccidiosis by eliminating the buildup of oocysts if any exists. You may also prevent the disease by using medicated starter as well as growth feeds.

Treatment

You will want to keep a solid batch of sulfa drugs and amprolium close by in case the disease strikes (which it does with alarming consistency). Administer these drugs in the drinking water of your chicken.

Mycoplasmosis

Symptoms

Your chicken will exhibit dirty nostrils, watery eyes, sneezing and bouts of coughing. Your chicken will also be slow in developing. Hatchability, fertility and production of eggs all register drops. Over time, cheesy material collects in both the sinuses and the eyelids. Noticeable outward swelling is also observed.

Prevention

You will find that the best prevention for MG is to purchase chicken that are free of the disease, if at all you get around to buying grown chicken rather than raising them from the chick stage. However, this is often more easily said than done given that the carriers often appear in perfect health. A basic and cheap blood test that is carried out by the bulk of veterinary diagnostic labs will detect any prior exposure to the disease. Start with contacting your veterinarian or local extension office and ask for an expert to help you.

Treatment

You will have to begin at the beginning here- lower any stress boosters that may be present. Clean the coops and reduce dust. Follow this up promptly with antibiotic treatment and proper nutrition. Both tetracycline and tylosin are some powerful antibiotics that will help do away with symptoms of the disease. However, if you are dealing with a carrier chicken, be warned that the antibiotics may not be able to do much as far as complete curing is concerned.

Administer antibiotics via the drinking water but be keen not to administer them for a period longer than 7 days. Vaccines are generally not encouraged in most states in the US, especially given that they tend to foster a milder version of the same disease they are supposed to help keep at bay.

Colibacillosis

Symptoms

Your chicken will appear disturbed and listless, show labored breathing and have their feathers almost permanently ruffled. They will also cough relentlessly. If the infection has been allowed to stew, your chicken will diarrhea, exhibit swelling as well as spleen and liver congestion. In the case of newly hatched chicken, navel infection may be observed.

Prevention

To cut down on the risk of infection, keep the area where your chicken live free of dust, feces and also ensure that there is sufficient living space for each of them. Do not use eggs that are visibly dirty for hatching purposes as these may well perpetuate the disease. Efforts to clean the eggs may lead to the eggshells cracking and this will open them up for increased bacterial penetration.

Treatment

The disease responds to antibiotic treatment. Especially where tetracycline and sulfa drugs are used, results are often positive. Make sure treatment exceeds 5 days before you evaluate any improvement in disease symptoms.

Key Point

Lowering stress conditions pops up in just about every preventative measure. Keep the dust away and the feces removed; make sure the drainage system is flawless and you will have made a huge step towards a disease-free chicken farm.

Different Chicken Breeds: What To Know About Varied Breeds Before Selection So As To Boost Your Chances Of Owning Strong And Healthy Chicken

"Selection is everything. By default, all the survivors of nature-that lot that has managed to outlive the rest and establish a line of offspring, were selected by mother earth to perpetuate the course. You too must learn to choose what to go for to truly succeed where you may."

-James May.

Here are some factual comments of different chicken breeds. While selection will ultimately come down to you, it will be a sapient thing to follow the recommendations given, where they appear.

Golden sex link

This breed of chicken is characterized by weak immune systems. This said, you should know that they are not the hardiest of breeds and are easily bullied by other chicken breeds. While you may perhaps have a different opinion on these, it is generally not recommended that you settle for these.

Source: Calranch.com

The Americana Breed

These are remarkably beautiful show birds. They are strange and moody though. These birds, while eye catching in their splendor, will not fit the profile of the utilitarian bird most people prefer. Especially, they will not fit the profile of birds that you would prefer to have on a working farm for the production of food.

Source: Imgkid.com

The leghorn

This breed is a phenomenal layer of eggs. If there is any fault to them, it is that they are generally distrustful of human folk. You are guaranteed to have some trouble catching these but when all is said and done, the eggs they lay will probably do a lot more for you than rewarding you for a tiring chase.

Source: Chickenbreedslist.com

The Delaware white breed

This bird is a great hybrid bird when it comes to producing eggs and providing the family with meat. However, in as much as they make a superb hybrid for both egg production and meat provision, they are not the absolute best for each one. Compared to other breeds, they grow to quite a large size and with this growth and girth, comes a remarkable appetite.

While their pure white appearance is really quite beautiful to behold, it ranks up there, along with the rest of the drawbacks package. Their pure white coat makes them easy prey for predators, as they are easy to spot from long distances. Birds that are more natural colored tend to blend in better with the natural terrain.

The barred rock breed

These birds are a favorite for many chicken farmers. It is not unnatural to hear a farmer reply to a barred rock-question with, "I like these birds. I just do". The reason for the love is that these birds are among the easiest to handle, are fairly productive and above all, are quite a hardy breed. This breed has been described as one of the most consistent picks of favorite breeds for farmers.

Source: imgbuddy.com

The Rhode Island Red

This breed is simply fantastic. It is, by a mile, the most potent survivor of all the breeds mentioned above. In the opinion of most experienced farmers in different parts of the country, no other breed comes as close to perfection as the Rhode Island Red, especially in a farm context. They are very friendly as well and are handled with ease. Here is a testimonial from one of the most renowned poultry bloggers in the country, "During the seasons of spring and summer, my Rhode Island flock will lay between 6 and 7 eggs per week, each."

This is quite remarkable in itself. Simply put, if you are a novice when it comes to chicken and are looking for a breed to kick things off with, just go with the Rhode Island Red breed. In addition to a supreme hardiness, friendliness, and easy handling, they are some of the most prolific layers around.

Source: Thechickentreet.com

Key Point

Hardiness, in addition to solid egg laying and meat production, is a key trait to look out for. Try to settle for breeds that show a tenacious will to survive.

How to Apply What You've Learned?

Perhaps, the best thing about this book is that while it does detail and caution against problems, it does not make out the practice of raising the "fair fowl" as nuclear science. The straightforward directives, instructions, and explanations will make it all the easier to apply what you learn.

Following what it details to the letter and succeeding while you try will require the support of an expert, at least in stretches. Call up a vet, perhaps one with whom you have enjoyed a professional relationship for some time, and ask as much info from them as possible. Above all, keep your chicken coop clean: this book will hardly help you if you do not keep the living places clean. If you have noticed, your success towards raising strong healthy chicken will depend on how clean you keep the coop.

www.ingramcontent.com/pod-product-compliance
Lightning Source LLC
Chambersburg PA
CBHW060403190526
45169CB00002B/736